SUNY
DOWNSTATE
Medical Center

Training Program to Enhance Cultural Competency in Nursing Homes

A MANUAL FOR TRAINERS

A collaborative effort between SUNY Downstate Medical Center and Horizon Care Center

Carl I. Cohen, MD ✶ Ummulkhair Muhammed, MA MS ✶ Monique Bowen, MA MPhil

Funded by a grant from the New York State Department of Health

A training program developed by

Carl I. Cohen, M.D., Director & Distinguished Service Professor
Ronald Brenner, M.D., Co-Director
Georges Casimir, M.D., Co-Director
Ummulkhair Muhammed, M.A., M.S., Project Administrative Officer
Monique S. Bowen, M.A., M.Phil.

SUNY Downstate Medical Center, Brooklyn, N.Y.

Nechama Markowitz, Administrator

Horizon Care Center, Far Rockaway, N.Y.

Zachary Cohen, Copy Editor

Funded by the New York State Department of Health

Copyright © 2012 by the New York State Department of Health

All Rights Reserved.

ISBN 978-1-105-26901-1

Acknowledgements

This manual was made possible by a grant from the New York State Department of Health *Dementia Grant Program* awarded to Horizon Care Center. We thank the following organizations and individuals who contributed to the development of this manual: Stephanie Heverly; Beth Dichter, Ph.D; Victoria Parker, EdM,; Hal Barton, PhD; Andreas Adams, EdD, MSW; Margaret Briggs, MBA; Beverly Collier, MS; Louis Ebarb; Ustinov Luke; C. Anthony McClaren; Joan McGowan; Sharon McKenzie-Reece, PhD, CTRS; Anne McNeill, MS; Michelle Melendez, MS, LCSW, CASAC; Glenda Perreira MBA; Barbara Singh; Lorna Walcott-Brown, MS; Sam Sanchez; Taphat Tawil; Barry Dukes; Ralph Zimmerman; Sara Cohen, M.D.; and Rhoda Meador. Special thanks to the administrators, staff, residents, and family members from the following facilities:

- Rutland Nursing Home
- Townhouse Center for Nursing and Rehabilitation
- Horizon Care Center
- CABS Nursing Home Company
- Center for Nursing and Rehabilitation
- Resort Nursing Home
- Grace Plaza Nursing and Rehabilitation Center
- Caton Park Nursing Home
- Saints Joachim & Anne Nursing and Rehabilitation Center
- Rockville Nursing and Rehabilitation Center
- Hempstead Park Nursing Home

For more information about the program, contact:

Carl I. Cohen, M.D.

SUNY Distinguished Service Professor

Director, Division of Geriatric Psychiatry

SUNY Downstate Medical Center

Box 1203

450 Clarkson Avenue

Brooklyn, N.Y. 11203

E-mail: carl.cohen@downstate.edu

TABLE OF CONTENTS

ABOUT THE PROGRAM 1

MODULE 1: INFORMATIONAL COMPETENCE 3

Description 3

Explanations of Significant Concepts in Cultural Competency 6

Health Beliefs, Values and Attitudes 10

Video Case Studies 21

MODULE 2: INTRA-PERSONAL (SELF-AWARENESS) COMPETENCE 24

Description 25

Video Case Studies 30

Cultural Sensitivity 32

MODULE 3: INTERPERSONAL COMPETENCE 38

Description 39

Video Case Studies 46

MODULE 4: INTELLECTUAL COMPETENCE, INTERVENTION COMPETENCE, COGNITIVE STRATEGIES 50

Description 51

Video Case Studies 60

APPENDICES 61

Test Your Knowledge on Cultural Competence 61

Frequently Asked Questions about Cultural Competency 66

Instructors' Post-Evaluation 69

RESOURCES 71

ABOUT THE PROGRAM

Welcome to the *Training Program to Enhance Cultural Competency in Nursing Homes*. This educational program was field tested in ten nursing homes in NYC. The themes in the program are based on information gathered from focus groups. Before turning to the actual training materials, it will be helpful for you to understand the conceptual basis of the program.

At one time, "cultural competence" was seen as simply addressing language obstacles and learning more about specific cultures. Although these are important, sometimes the way we describe a specific culture can create misrepresentations or stereotypes of people. Cross-cultural training is now defined as teaching health providers to understand, communicate with, and provide quality care to patients from different backgrounds (Weissman et al. 2005). It is really part of what is called "patient-centered care." Patient–centered care means that health providers respond to patients' needs, values, and preferences.

Therefore, the model for the way we think of cultural competency has three elements. Health providers must: (1) be warm, understand others' feelings, and be genuine; (2) respect people who are culturally different from themselves with respect to their needs, values, and preferences; and (3) learn skills for interacting with persons who have cultural needs, values and preferences that differ from their own.

The *Training Program to Enhance Cultural Competency in Nursing Homes* is designed to improve staff members' understanding of residents' cultures as well as their individual needs. When staff is more knowledgeable and accepting of residents' needs, it is more likely that staff and family members will work cooperatively to improve the quality of care for residents.

This manual is a guide designed to accompany a four-module video program. The video can be used effectively in one-on-one settings or in small group settings. Each module presents learning objectives and contains images from the video to further highlight the theme in the module.

All modules are independent. However, the module sequence we suggest is designed for optimum learning about cultural competency. Participants are encouraged to engage in small group discussions after viewing the video. In the video, a voiceover is presented to stimulate discussion and promote individual reflection. Group activities are also described following each video so as to provide additional learning opportunities.

The aim of this curriculum is to conduct a cultural competency training program for nursing homes. In so doing, it will focus on three patient populations: African Americans, Latinos, and Caucasians. We include persons of Jewish ancestry, who represent a large portion of the Caucasian residents in nursing homes in New York City, so as to illustrate certain themes with respect to ethnic groups within the broader Caucasian population.

The training program consists of a video that includes four modules. The video is about 40 minutes long and can be used for individual learning, while the manual can also be a resource for the independent learner. Each module is conducted during a 45-minute session to allow for viewing the respective module and for follow-up group discussion. Each of the modules listed below will be covered in detail in this training manual:

Module 1 – Informational Competence

Module 2 – Intra-personal (Self-Awareness) Competence

Module 3 – Interpersonal Competence

Module 4 – Intellectual Competence, Intervention Competence, and Cognitive Strategies

MODULE 1:
INFORMATIONAL COMPETENCE

Description

This first module in Cultural Competency introduces basic concepts in the culturally competent care of residents in nursing homes. It summarizes information about history, cultures, and traditions as they relate to health beliefs, as well as stereotypes and generalizations. Ideas for increasing participants' understanding of residents as unique individuals are presented. While this section can be used with any cultural group, we have focused on persons of African heritage, Latinos, and Caucasians, mainly of Jewish ancestry.

Module 1 Learning Objectives

After completion of this module, participants will be able to

1. Define the major terms used in the cultural competency model.
2. Describe the qualities of health belief systems.
3. Identify generalizations and ways to dispel stereotypes of Blacks, Latinos, and Jews.

Explanations of Significant Concepts in Cultural Competency

Culture includes common values, norms, traditions, customs, art, history, religious and spiritual practices, and institutions of a group of people. It shapes personal and group values and attitudes. Culture helps explain how persons view and act within their social worlds.

Cultural Competence is a skill that allows individuals to increase their understanding and appreciation of cultural differences and similarities within, among, and between groups and individuals. Knowledge of diverse cultures is often necessary to communicate with understanding to members of a different culture.

Cultural Sensitivity is having respect for customs and cultural values different from one's own. Cultural sensitivity is seen as the first step towards cultural competency.

An **Ethnic Group** is a cluster of individuals with a common sense of uniqueness based on race, religion, or place of birth. The members of the group express a similar self-identity by having a common heritage and unique social characteristics. Sometimes, the behaviors of the group are not fully understood by outsider.

Ethnocentrism is belief or attitude that one's own cultural view is the only correct and best view. This belief is often held unconsciously and grows out of the assumption that one's own culture is the "norm."

Minority is a smaller group within a larger group that is often perceived as a group in need. It is often used to name a group that experiences discrimination and, as a result, its members may feel that they are "second-rate" because of their race, ethnicity, or place of birth.

Race is the categorization of a group of people based on their shared visible characteristics. The term is somewhat problematic because of the widespread diversity among racial groupings. (Note: Although the term race has been used for a long time to distinguish groups based on heritage and physical differences, its use is controversial because of the lack of specificity and scientific support for racial distinctions.)

Spirituality is the expression of one's faith or religion, but can also refer to the beliefs and values tied to the importance of nature, of life energy forces, of goodness in all things, or of family and community.

Impact of Cultural Factors on Competent Care in Nursing Homes

Our "Western" health care system has its own culture: knowledge, beliefs, skills, and values based on scientific assumptions and processes. Modern medicine contains certain beliefs that are used to define and explain diseases. Residents in nursing homes may identify cultural factors or beliefs that do not match those found in modern medicine. However, all of these factors can have a direct impact on health care, recommended treatments, and communication between residents and their care providers. Residents may define physical disorders based on their own cultural beliefs and they may have their own opinions about treatment. At times, a conflict between modern medicine and a resident's own belief system can arise and make the practice of culturally competent care particularly difficult.

Because there may be cultural and language differences between a care provider and a resident, the care provider may need to take additional steps to reduce confusion and misunderstanding. In some communities of African and Latin descent, there can be a general inclination to answer "yes" to a question posed regardless of whether or not the question is fully understood. The "yes" response can be interpreted as an act of respect toward the treating clinician. In this example, the simple technique of asking the residents to explain in their own words what they understood can provide important information. If a care provider does not speak the same language as the resident, a trained interpreter must be used. A *translator phone* is another option to improve communication with non-English speaking residents.

With respect to medication and surgery, cultural differences in attitudes and beliefs can also be different. Some residents may willingly take medication, whereas others, because of cultural beliefs, feel that medications such as antidepressants are harmful, dangerous, and indicative of mental illness.

Care providers should work diligently to understand the key aspects of the culture and behavior of the residents. In order to provide culturally competent services to residents, care providers must value the differences and similarities in health beliefs, remain open to learning about culture, and treat all residents in a respectful manner.

Ethnocentrism as a Negative Consequence

Like all people, care providers may interpret cultural situations different from their own by using their personal beliefs to evaluate the situation. This is an example of *ethnocentrism*. It often leads to miscommunication, which can lead to stereotyping and disrespectful interactions. Once this occurs, it will most certainly result in a breakdown of the provision of culturally competent care. Furthermore, if care providers do not have experience working with individuals from different cultural backgrounds, they are more likely to prejudge those in their care based on stereotypes, hearsay, and emotional reactivity. These judgments may lead care providers to overlook a unique quality of a resident's culture, such as his or her language, beliefs, habits, or behavior. The result may be an offended resident, who is then potentially less likely to follow treatment recommendations.

Health Beliefs, Values, and Attitudes

Medical Pluralism in the United States

Elderly residents from any one ethnic background may or may not know, or may not take on, the health beliefs connected with their cultural heritage. Because there are many health beliefs in America, residents may borrow from two or more cultures. Acculturation is taking on the main culture in society. For instance, a person of Latino background who does not speak Spanish is considered highly acculturated. Religion is another cultural factor that may influence health beliefs and choices regarding treatment. For example, Jehovah's Witnesses place a heavy emphasis on the healing power of God and blood is seen as sacred, and they may not accept blood transfusions.

The descriptions in this module of the health belief systems of people of African, Asian, and Caucasian descent are generalizations of each group. In fact, within each of these larger racial groups, smaller ethnic communities exist that have very different beliefs and ideas from others within their own racial grouping(s). *At no point do we want these differences to be used to encourage continued stereotyping.*

The historical experiences of many ethnic groups with respect to immigration and discrimination provide information about how these groups view themselves in the context of the broader American culture, and can explain some of how they think about health and health issues. Furthermore, historical events affect residents differently at different points in time: one's reaction to a past event can vary considerably based on the age of the resident now, versus their age at the time of the event, or as compared to the age when the person learned of the event that had an impact on his or her ethnic group. It is important to note that not all residents who identify themselves as members of an ethnic or racial group share the same experiences, or have an awareness of others experiences. Asking residents to share their social histories, and to listen to stories of other people's experience, can help them to understand themselves and others better.

Health Belief Systems influenced by Western Values

Caucasians in the West often place great importance on science to solve medical problems. This value system has influenced the design of hospitals and nursing homes in the United States. Many values that have influenced this belief system are from Europe, the Middle East, and Ancient Greece. Technology, material things, and the human body are often considered to be more important than the soul or spirit.

The immediate family (also called the nuclear family) is often considered to be the ideal family structure. Families often have one person as the head of the family. Sometimes family roles can be equal. There is great importance on setting goals with regard to education, career, and retirement. People who are influenced by Caucasian values may be more likely to:

1. **Make decisions about end-of-life issues or complete advance directives.**
2. **Agree to withhold "life support" treatment.**
3. **Agree to withdraw "life support" treatment.**
4. **Use hospice services.**
5. **Agree with physician-assisted death.**
6. **Be willing to donate organs.**

Health Belief Systems influenced by African Traditions

Many African American elders from the farming areas in the southern United States were once accustomed to providing medical treatment to themselves and others. Prior to the 1960s, Black Americans could not get treatment in many of the places that their White counterparts received care because of legal discrimination and segregation.  Along with an abiding legacy of religious steadfastness, African Americans from older generations have generally maintained an allegiance to Christian values, with churches figuring prominently within many African American communities. In some North American communities, and to a greater extent in some Caribbean island nations, the spirit world, local shamans, and family talismans are of significance and may be called into use to treat minor to more substantive illnesses.

As with other world cultures, family and familial connections are typically very important to people of African descent, with grandparents and other elderly members of the community viewed as an integral part of a comprehensive, kinship network. Here, even people who are not blood relatives but share a close bond may be called "auntie" and "uncle" as a sign of respect for their age, experience, or role within a community. Thus, people influenced by cultural traditions of filial piety, may believe that:

1. **Power and wisdom reside in commonly-held ideas and traditions, and that such beliefs and practices can cure medical problems.**

2. **Natural substances (e.g., roots, herbs) can be as much, or more therapeutic than man-made and mass-produced medicines.**
3. **One's faith, spirit, and soulful connection to one's ancestors can heal the sick.**
4. **Paying attention to the present rather than dwelling on the past is a preferred path to physical and mental fortitude.**

Health Belief Systems influenced by Latino Traditions

Many ethnic traditions are influenced by a mixture of cultural beliefs and by the practices of transplanted people from around the world. The impact of cross-cultural influences can be observed in countries throughout Africa, Asia, Europe, and The Americas, particularly in the Caribbean and in coastal locations throughout Central and South America.  Here, languages, customs, and religions converged to create many Latin American sub-cultures with multiple variations on similar themes, such as the importance of family, faith, ancestral linkages, morality, and the value of living things. Thus, it is common for Latino families—like so many other cultures worldwide—to emphasize the roles of certain family members.

Elderly people are revered and typically held in high esteem as being wise and experienced. As a result, elders are often consulted when a family has to make key decisions. Although some families may refer to the eldest male in a unit as the "head," or the lead, when important matters come up, it is more typical that the leader has consulted directly with other elders in the family before making a decision. Although, in some cases, a male family member may be viewed as the head of the family to the outside world, it is more typical for the eldest family member—regardless of their sex— to serve as the lead role model within the family of origin.

Religious influences for Latinos include native/indigenous American traditions, Christianity (mainly Catholicism and Pentecostalism), and African religious customs. Many religious Latinos believe in saints and regardless of specific religion may observe a patron saint day in observance with the prevailing community norms. The use of religious symbols to keep good health, as well as herbal remedies is common practice among many older Latinos, and is one of many values that are passed down in families. Furthermore, there is a transgenerational belief in native healers (e.g., curanderos/as, "santero/as", espiritistas, "vodou" priest and priestess), with great importance on spirituality in the various forms that may take. People who are influenced by Latino traditions may be more likely to:

1. **Hold strong beliefs about the importance of closeness with family and other close relatives.**
2. **Place great significance on their spirituality.**
3. **Have a strong sense of community.**
4. **Value both the past and the present.**

Stereotypes and Discriminatory Generalizations about Ethnic Groups

A stereotype is a negative, oversimplified generalization used to describe a person using a group definition rather than a set of unique characteristics. The list of stereotypes is endless and usually hurtful. They can be perpetuated through visual images in print, television, motion pictures, or internet websites. Because the human brain naturally classifies information by type, our capacity to note differences among people by their physical traits first occurs outside of our conscious awareness. This type of grouping initially develops unconsciously. Stereotyping may develop from our direct or indirect experiences with others, along with other influences that are by-products of interactions occurring within the social environment. Forming a stereotype is not abnormal or morally wrong. However, basing our judgments and actions on a widely held but oversimplified conception of a person or group of people, has been and continues to be the root of many social problems.

People make judgments: we make them about people based on our experiences and commonly held beliefs. Often, we are not fully aware that we use our experiences and beliefs to shape our ideas about how people of the same group will behave or interact with others. So, for example, care providers who mainly perceive another ethnic group's differences ahead of any shared similarities may call on negative images and ideas that can inhibit them from approaching a resident with an open mind. As a result of biases, care providers neglect the resident's strengths and unique qualities, and may miss opportunities to make appropriate interventions. Staff must always remember that a resident is an individual first. Reflecting on a resident's ethnic group can occur in ways that are sensitive and respectful of cultural differences. The best results include care providers moving beyond preconceived ideas about the residents in their care, instead basing their work with people on their knowledge of those individuals within the context of multiple cultural determinants.

In sum, a stereotype, or a generalization, is a shorthand description of a person or a group using broad definitions rather than unique characteristics. The list of stereotypes that people use is endless and usually hurtful, and is not supported by the fact that there is great diversity among groups of people. The list on the next page provides some responses that can be used to counteract generalizations and to limit the harmful impact of stereotypes.

A Tool to Debunk Stereotypes

Ethnic Group	*Stereotype*	*An example to counter the stereotype*
Jews	Don't like to spend money; not generally recognized as inventors	**Sidney Kimmel**, Founder of Jones Apparel Group, Inc. Supports cancer research, the arts and Jewish continuity donated $341,000,000 from 2000- 2004. **Charles Ginsburg** invented the videotape in 1950s
African Americans	Not scholarly; have little education	**Ben Carson,** born 1951, is currently Director of Pediatric Neurosurgery at Johns Hopkins Medical Institutions **Marie Maynard Daly,** born in 1921 is the first black American woman with a Ph.D in chemistry.
Latinos	Don't like to work hard; involved with drugs	**Baruj Cenacerraf**, Venezuelan, Nobel Prize for medicine and physiology. **Carlos Finlay,** Cuban, identified the mosquito as a carrier of the deadly yellow fever.
If time permits, participants can fill in columns with their own examples.		

Sources

Ho, M.K., Rasheed, J.M., Rashedd, M.N. (2004). *Family Therapy with Ethnic Minorities*. 2nd Edition. Thousand Oaks, Ca, Sage Publications.

Yeo, G., Hagan, J., Levkoff, S., Mackenzie, E., Mendez, J., Tumosa, N., Wallace, S. (1999). *Core Curriculum in Ethnogeriatrics*, Developed by the Members of the Collaborative on Ethnogeriatric Education. Bureau of Health Professions, Health Resources and Services Administration, U. S. Department of Health and Human Services.

Video Case Studies

The case studies in this manual are provided to help training participants learn about themes covered in each module. The voice-overs in the video promote self-reflection and group discussion. Each section is designed as a tool to prompt discussion. After the course instructor plays the video for participants, s/he will facilitate a group discussion about the case studies depicted. For these group sessions, activities are provided following the case descriptions to help participants reflect and learn from the videos. This video can also be of value when used independently by individual care providers.

Case 1. Health Beliefs

The clip below introduces the concept of disparate cultural health beliefs. An Afro-Caribbean-born resident expresses his views on his illness and treatment. His Caucasian-American care provider offers a different view. In the video, voice-overs present other alternatives.

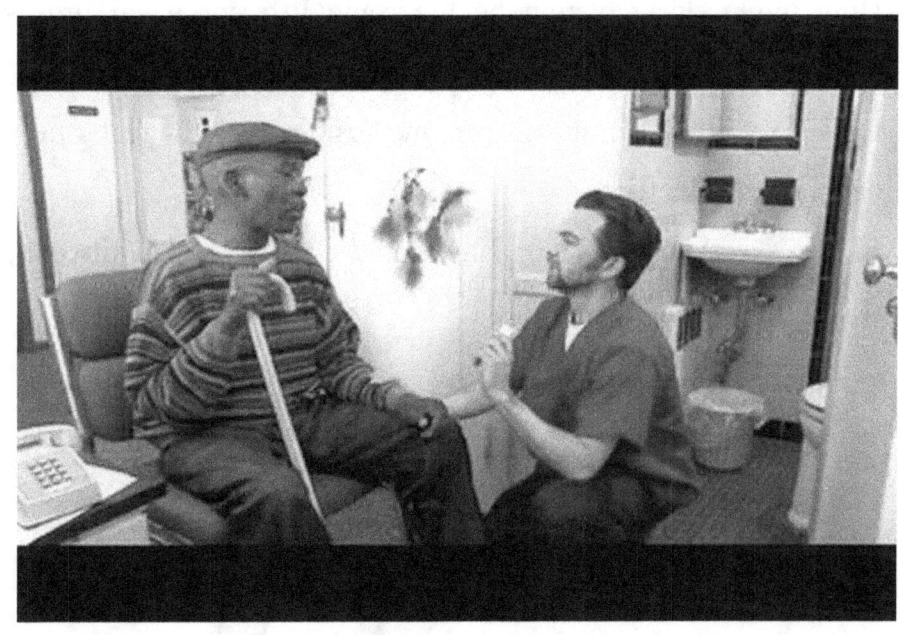

Case 2. Health Beliefs

This clip shows a Latina resident and her family member conversing in Spanish. Her Caucasian-American, English-speaking care providers cannot understand what is being discussed and express their suspicions about the content of the conversation.

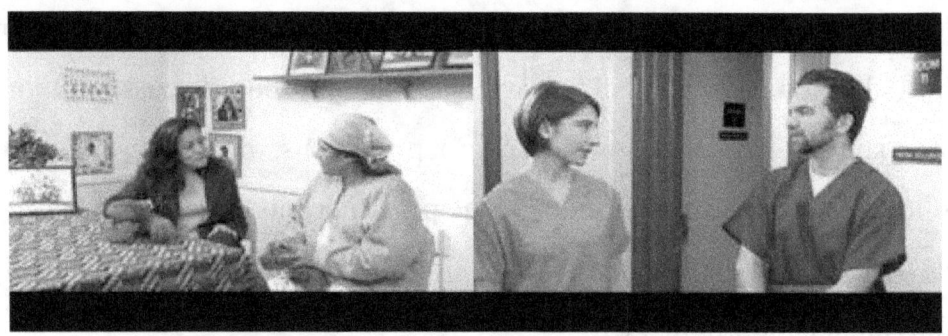

Activity: Reflecting on Stereotypes

- What is happening for the staff member? Have you ever been in a similar situation?
- How can a staff member become more comfortable learning about and providing care for this resident?
- How might this situation be the same/different when it is staff members that are speaking in a language other than English?
- How do you think English-speaking residents might feel when this happens?
- In the case scenario, is there anything the resident or family member should do to help?

Refer to the voice-overs from the video for ideas to help facilitate discussion relating to how these activities might help reduce stereotypes.

Activity: Stereotypes of Ethnic Groups

- Have small groups of three gather to share what they think or know about Blacks, Latinos, and Caucasians of Jewish ancestry, or another ethnic group at your residence.
- Participants should ask themselves and fellow members, "How do I know that?" "Is there factual evidence to support this?" "What are different examples that exist that are contrary to my experience?" (See prior and upcoming sections for "Debunking Stereotypes" and "Strategies to Reduce Stereotypes".)
- In the large group, discuss how our ideas about a particular ethno-cultural group are formed, and the ways that these notions are further influenced by the media.

Refer to the voice-overs from the video for ideas to help facilitate discussion and on how these activities might help reduce stereotypes.

Activity: Strategies to Reduce Stereotypes

- Engage in self-reflection and introspection
- Attend training sessions in multiculturalism
- Visit museums and historical sites that feature various ethnic groups
- Participate in inter-group gatherings
- Read/research cultural groups from a variety of sources
- Seek out movies that raise awareness about prejudice

Refer to the voice-overs from the video for ideas to help facilitate discussion.

MODULE 2: INTRA-PERSONAL (SELF-AWARENESS) COMPETENCE

Description

This second module provides activities that help participants develop a greater understanding of intra-personal competence. Care providers' capacity to reflect on their own thoughts and ideas is one of the major ways they can rate their self-awareness. The video in Case 3 is designed to help participants develop greater self-understanding about cultural stereotypes that they may hold.

Learning Objectives

After completion of this module, participants will be able to:

1. Explain their own personal beliefs and values and how this influences the stereotypes that they have about others.

2. Identify strategies and assessments to evaluate their own attitudes toward residents.

Intra-personal (Self-Awareness) Competence

Intra-personal competence refers to people's ability to become aware of their beliefs and behaviors, and to use this awareness to increase their cultural competence. Residents and care providers report that when the care provider is aware of how his or her actions affect others, residents experience greater satisfaction with care provider services. Furthermore, care providers with strong intra-personal competence have more productive professional relationships with recipients of care. In turn, residents show increased levels of respect toward these care providers and are more agreeable to treatment recommendations. Intra-personal competence also affects a care provider's level of performance: the capacity for **self-**

reflection is one of the principal ways a care provider can be rated on his or her ability to provide culturally competent care.

Developing Self-Awareness

Development of self-awareness can be used to guide each care provider in the self-assessment of feelings of prejudice and of discrimination. The first strategy care providers can use to strengthen their own personal traits is in being willing to listen to their own inner voices when they are feeling judgmental. Often, it is this inner voice that pushes to the heart of the issues and steers the care providers to find helpful rather than harmful solutions when having conflicting feelings. Another aspect of developing self-awareness is one's willingness to be open to and allow for different views. Genuineness encourages honesty and communication. Care providers who are psychologically closed off wear their insincerity in visible ways that demonstrate to others that they fear new ideas, and that they have a difficult time accepting that the perspectives of others.

It is also very important for care providers to be attuned to their own feelings of empathy, or lack of thereof, toward recipients of care. For instance, caring for residents, especially those with debilitating cognitive changes like those seen in advanced dementia, is a complicated task. Very often the behaviors of residents are frustrating to deal with. In such a situation, it is easy to feel overwhelmed—both physically and emotionally—and anxious when in role of the care provider. The question then becomes, "How does the care provider cope?" In this case, the providers working with residents with dementia must be trained to be patient and to be thoughtful about effective communication styles. Here are some strategies that can assist the care provider in being more effective and culturally sensitive:

- **Develop and maintain a sense of humor.** The art of providing good care involves maintaining a sense of humor and striving to "lighten up" about life's challenges. Describing a light moment from a movie or a program on television, sharing a funny story, or simply laughing with residents can help maintain positive relationships.

- **Seek out someone with whom you can confide.** A trusted friend, colleague, or counselor can make all the difference. You may need someone with whom you can talk things over, someone with some distance from the situation that will be nonjudgmental, be respectful of confidentiality, and be understanding of your needs.

- **Practice assertiveness and sensitivity.** It is a challenge to express one's own feelings and yet support the needs of others. Care providers may find that balancing multiple perspectives can inhibit them and lead to less effective communication. An important practice is assertiveness—that is, not being afraid to speak to family members, friends, and colleagues about your thoughts, feelings, and workplace needs. It is okay to admit that one does not always feel okay after a difficult interaction with a resident or with a family member. It is critical to be able to act assertively without being belligerent, aggressive, or passive-aggressive.

- **Organize a support group comprised of other professionals.** Many people find emotional support and even new friends through support groups. After all, sharing your experiences with others familiar with your profession can be satisfying, rewarding, and career affirming.

- **Keep a diary or notes of work experiences.** This can help care providers to problem solve. A diary can provide a safe place to write about your stressors and your feelings—good and bad.

- **Meditate**. Seek out an environment free from distracting, external stimuli. Find yourself a comfortable place to sit quietly, with or without shoes. Whenever possible, seek calm and silence. Silence is one of the most powerful counterbalances to anxiety.
- **Document and inform as a way of seeking support.** Letting your supervisor know about challenges that you face can be an excellent way to record your experiences formally, while also soliciting help in making positive workplace change(s). Supervisors can play a vital role in providing support to staff; however, sometimes supervisees have to be prepared to ask for the type of support that feels most effective to them.

As a care provider, it is of the utmost importance to stay in touch with your own feelings. The difficult behaviors displayed by persons with dementia often can cause care providers to feel frustrated, angry, upset, and stressed. It is *okay* to feel this way. However, it is *not okay* to express your negative feelings in a destructive way. You can control and cope with your negative feelings, but the resident you are caring for likely cannot modulate their feelings very well. For instance, yelling at someone with Alzheimer's disease is likely to make the situation worse for everyone. Remember that the resident's behavior is not deliberate. If you berate a resident in response to an unwanted behavior, it will still be considered abusive. One way of dealing with a difficult situation like this one would be to separate from it and the people involved, as soon as possible. Next, use the strategies that were discussed earlier to bring about a sense of inner calm. It is important *to practice* the techniques often so that they become automatic.

Source

Adams, A. & Walcott-Brown, L.(2001), *Alzheimer's Training Curriculum for Direct Health Care Providers,* Brooklyn Alzheimer's Disease Assistance Center (BADAC), SUNY Downstate Medical Center, Brooklyn New York.

Video Case Studies

Case 3. Self-Awareness

Show the video and listen to the voice-overs. The video clip depicts a scenario that is designed to illustrate the impact of negative stereotypes. In the clip, two care providers reflect on their own stereotypical views about persons of Jewish ancestry. After you show the video clip, facilitate the following activities with participants.

Activity: Self-Awareness

- Ask participants to respond to the following question.
- What resident behaviors make me feel uncomfortable?
- How do I respond when I feel uncomfortable?
- What are my own biases and assumptions about the residents in my care?
- Do I consider other people's ideas as valid and valuable as my own?
- When I judge others, how do I feel and what do I do with these feelings?

Refer to the voice-overs from the video for ideas to help facilitate discussion relating to this activity.

Cultural Sensitivity

Cultural competency in nursing home professions encourages the integration of conventional Western approaches with the cultural health  beliefs of the residents and their families. Not without challenges, this goal is best achieved when the care provider maintains a culturally sensitive point of view. Without an appreciation for the power dynamics involved in cross-cultural interactions with recipients of care, health providers may allow for self-oriented cultural perspectives to dominate and influence their interactions with recipients of care. Any lack of awareness of the potential impacts could result in breakdowns in communication between residents and care providers.

A Developmental Model of Ethnosensitivity or Cultural Sensitivity

The behavioral responses represented herein explore care providers' and residents' experiences and how these interactions impact caring:

Fear is the highest level of cultural insensitivity. In order to reduce fear and uneasiness about differences among ethnic groups, one must place less importance on one's own opinions and attitudes. It is important to understand that all views regarding culture and traditions are of equal value.

Denial of racial and ethno-cultural prejudices is the refusal to acknowledge the ways in which individual differences can be used to discriminate against minority groups.

Superiority breeds the dangerous belief that one group's identity and attributes are better than others. Cultural sensitivity means removing negative labels or stereotypes in society and encouraging respect and recognition of similarities and differences.

Minimization is a common way that people attempt to reduce feelings of guilt associated with unethical behavior. It also encompasses "cognitive distortion," wherein one attempts to downplay the importance or the impact of events that bring about negative emotions, or attempting to minimize the perception of an impact of another's behavior on one's self.

Relativism is the concept that there is no absolute truth or validity to one's viewpoint(s). Despite inherent cultural biases within all people, it is when one can accept his own perceptions as different from someone else's and yet embrace rather than reject another's viewpoints that are real understanding and compassion for all people becomes possible.

Empathy is the capacity to recognize and share feelings that are experienced by another person or culture. The ability to respond to all cultures as of value is an important foundation for our interactions with each other.

Integration is the highest form of cultural sensitivity. Persons who hold this view use their skills to promote a multicultural environment where all persons are respected and valued.

Source

Borkan, J., Neher, J. (1991). *A Developmental Model of Ethnosensitivity in Family Practice Training. Family Medicine*, 23:212-217.

Activity: Cultural Sensitivity

> In many ethnic cultures there is great appreciation of elders. Care providers should show appropriate respect to older residents. Some strategies to show respect to elderly persons include
> - Offer an appropriate greeting, even if there are language differences.
> - Acknowledge the resident in a personal way before asking business related questions. For example, within the Mexican culture it is customary to ask "How is your family?" when establishing rapport.
> - Avoid the "Invisible Resident Syndrome." Talk <u>to</u> residents rather than <u>about</u> them when in their presence. Otherwise, you may be seen as disrespectful.
> - When attending to a resident, be sure to use their name.
> - Use appropriate cultural gestures to greet the person. In some cultures the failure to shake hands is hurtful.
>
> *Refer to the voice-overs from the video for ideas on facilitating discussion related to this activity.*

Activity: Values Clarification

> - We are all "ethnic" in some way. This exercise requires that you consider aspects of your own family background that concern ethnicity.
> - Reflect on experiences you or other family members have participated in or that you enjoyed specifically because of your family's ethnic background, racial identity, or class background. Can you name any specific activities, or privileges and advantages that were common to you and your family, but perhaps would seem rare or unusual for someone from another group to have firsthand experience doing?
> - List all the strengths or advantages.
>
> *Refer to the voice-overs from the video for ideas to help facilitate discussion relating to this activity.*

Positive Feedback to Build Relationships

Have you observed that some people have problems complimenting others, and yet there seem to be those who have difficulty receiving a compliment? Although there may be many reasons to offer a compliment, some people say nothing--keeping the thought as a private one--while others try to be sure to say something kind as a general way of interacting with others. You may find that you and some care providers you know may behave in a similar way. Regardless of where you fall along the continuum, one thing is certain: Letting another person know you are thinking about them and something positive they may have done or said goes a long way in establishing and maintaining rapport.

So, in a long-term care setting, an effective communication strategy would be to offer a positive statement as an acknowledgement of a desired behavior. Furthermore, rather than remarking only when a resident does something negative, try also to comment on the occasions when the resident is doing something positive, and that praise can be offered to the person as a compliment. Praise can serve as that reinforcement of that behavior you want to encourage in your residents. As with most things, moderation is the key. It is easy to appear disingenuous because praise is offered at every turn. Instead, provide thoughtful and stable encouragement at well-timed intervals to achieve the best results.

Activity: Giving and Receiving Praise

> - Have each participant turn to the person sitting beside them and pay him/her a compliment about something he/she has said throughout the exercise(s). The acknowledgement could also be personal (if you have relationship with the person where that would not be too unusual) or about behavior. Ask participants to think about how it feels to send and receive praise.
>
> *Refer to the voice-overs from the video for ideas on how to facilitate discussion related to this activity.*

MODULE 3: INTERPERSONAL COMPETENCE

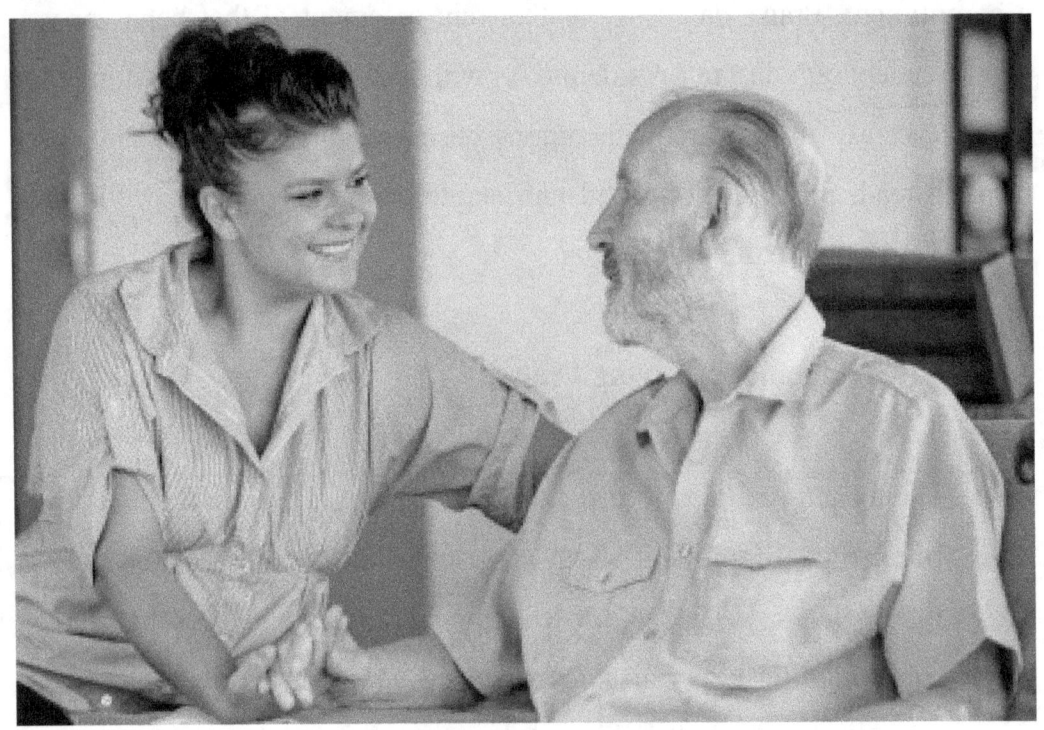

Description

This module is designed to help care providers learn to improve their communication skills with residents by using both verbal and non-verbal strategies.

Learning Objectives

After completion of this module, participants will be able to:
1. Describe ways to express understanding, respect, and warmth.
2. List approaches that help to promote good communication between care providers and residents.

Compassionate and Sensitive Communication in Nursing Homes

Good communication includes the successful flow of information from one person to another. The listener must receive the information in order for communication to have occurred. Moreover, for information transfer to be done well, the message sent has to be in sync with the one received. For the most part, this is not always easily accomplished. There are many points along a lit path where the exchange of information can become lost, obscured, or misunderstood.

The ability to communicate effectively is a desirable trait for care providers. Positive care provider/resident relationships develop when there is an ebb and flow style of communication. These skills are improved upon when care providers engage in a regular effort to use empathetic and active listening strategies.

Understanding of verbal and non-verbal communication includes knowledge and application of basic skills, such as using clear, simple language when speaking, and using active listening technique when working with others. Although communicating is something we all do daily, it can actually be quite complicated in the nursing home setting. Care providers must be able to communicate effectively to residents why certain tasks need to be done, and they may have to coax some residents to help them with the completion of other tasks. It takes consistent practice of listening with concern and compassion to acquire the necessary skills and, in turn, achieve the best outcomes.

There are three aspects to communication with the aging population in long term care settings that this module addresses: Effective communication with elderly residents; the impact of positive feedback on provider/resident relations; and the importance of active listening in the treatment of problems associated with aging.

How a Message is Received

Many factors can affect how a message is received. Hearing loss, medications, disabilities, and depression can have major affects to what extent residents can engage in mutually beneficial communication with the people around them. Elderly clients with hearing impairments are particularly vulnerable in residential settings and may need special attention from clinical staff.

Another reason a resident may not receive a message correctly is because the care provider may not be a good listener. A care provider may show inattention in several ways. In addition to not paying attention to the resident, either forms of inattention include care providers answering his or her own questions before residents finishes their responses; rushing or interrupting the residents; or offering advice before being asked to give it. Some of the reasons that care providers may not be paying careful attention to a resident are:

- They believe they already know what the resident is asking.
- They may have biases or stereotypes that stand in the way.
- There are distractions in the work environment.
- They may be suffering from fatigue or illness.

Care providers' tone of voice can also make an enormous difference in the way residents receive what is said to them, and can affect whether the resident feels adequately cared for. As discussed earlier, the process of communication is complicated, and there are many points at which the central message can be lost. Thus, it is important that care providers take enough time with their residents to improve how they respond to one another. It is especially important that the care provider take time to consider a resident's health-related issues through a cultural lens. As in acute care settings, many mistakes occur in facilities when care providers assume they know what residents are saying to them. Without "checking in" at appropriate intervals throughout a conversation, you fail to demonstrate active listening. Research has shown that when care providers check the content of what they believe they have heard, restate the message received in their own words as a check on their own understanding, and repeat this process across multiple conversations, then residents will (1) develop a sense that what they have to say is valued by the listeners (in this case, their care providers), and (2) feel increased trust in the providers of care.

Useful Listening Skills

A care provider who is an effective listener shows positive behaviors. Paying attention to the resident, adopting an accepting attitude, and allowing the resident plenty of time to speak are examples of effective listening. Conversations are most mutually beneficial and productive when a care provider stays on the subject, remains aware of his/her emotional reactions, and acknowledges his/her own cultural biases without letting those biases convert to judgments and prejudices.

Strategies to Improve Active Listening Skills

1. **Rephrase What Has Been Said**

Say again, in your own words, what the other person has said. This provides you an opportunity to check whether your understanding of what has been said is accurate. It also gives the speaker a chance to correct you if you have misunderstood them. For example, you might say, "So, I hear you saying…", "If I understand you, you are saying…."

2. **Echo The Speaker's Feelings**

The emotion that underlies the words is often more important than the words themselves. It is important to try to understand the strong feelings, attitudes, deeply held beliefs, and values. To convey that you heard what the speaker has said, you might say, "That must have hurt you." "I can picture that you are excited about that." "When that happened, did it bother you?"

3. **Request More Information**

In most circumstance, we need more information to understand what is being said to us, and sometimes we may need to ask for more explanations. Appropriate queries may include "I'm curious, could you please tell me more about that?" "What happened next?" "How did you feel when that happened?"

4. **Nonverbal Communication**

There are many ways in which people can observe whether we are listening and paying attention. Through our eye movements, in our body positioning, and by any manner of physical and vocal gestures we communicate how interested we are in what and involved we are in conversations. Next time you are talking to someone, try to observe what you (and the other person) convey nonverbally.

Moving from Thoughts to Words

Like anyone else, care providers can run into problems putting their thoughts into words. There are occasions when those thoughts and feelings are things that their residents should be made aware of. Although expressing feelings about a hurtful remark made by a resident is difficult, especially when one's culture, race, values, or traditions are at issue, having a respectful and an open dialogue allows the resident an opportunity to gain a better understanding of the care provider's experience. When the communication between the resident and care provider is strained, the provider may feel it is simpler to avoid the conversation. However, the research on building resident/provider relationships shows that providing firsthand information in a culturally informed manner is more effective than avoidance of sensitive subjects. In most cases, when one says, "I'm hurt," the usual response by an individual is, "I'm sorry." Revealing information about one's feelings can reduce conflicts and negative feelings.

The care provider has the opportunity to build trust and rapport with residents through day-to-day contact and conversation. The provider can encourage residents to share their feelings, especially about illness or loss. When talking with residents, it is important to discuss their concerns without making judgments. Residents should be approached in the least disturbing way. No matter the situation, care providers are expected to offer consistent and thoughtful support to all residents. If the care provider has any concerns about the resident, he/she should share those concerns with a supervisor or with another member of the care team. As health care professionals know, using a confrontational approach with a resident can be considered to be abuse.

Source

Cervantes, E., Heid-Grubman, J., Schuerman, C.K.(1995). *The Paraprofessional in Home Health and Long-Term Care. Training Modules for Working with Older Adults*. The Center for Applied Gerontology, Chicago, Health Professional Press.

Video Case Studies

Case 4. Racial Bias

In this clip, a white resident and a black care provider clash. The exchange is hurtful and shows a lack of understanding and respect based on prejudice. In the video, voice-overs present positive options toward resolution.

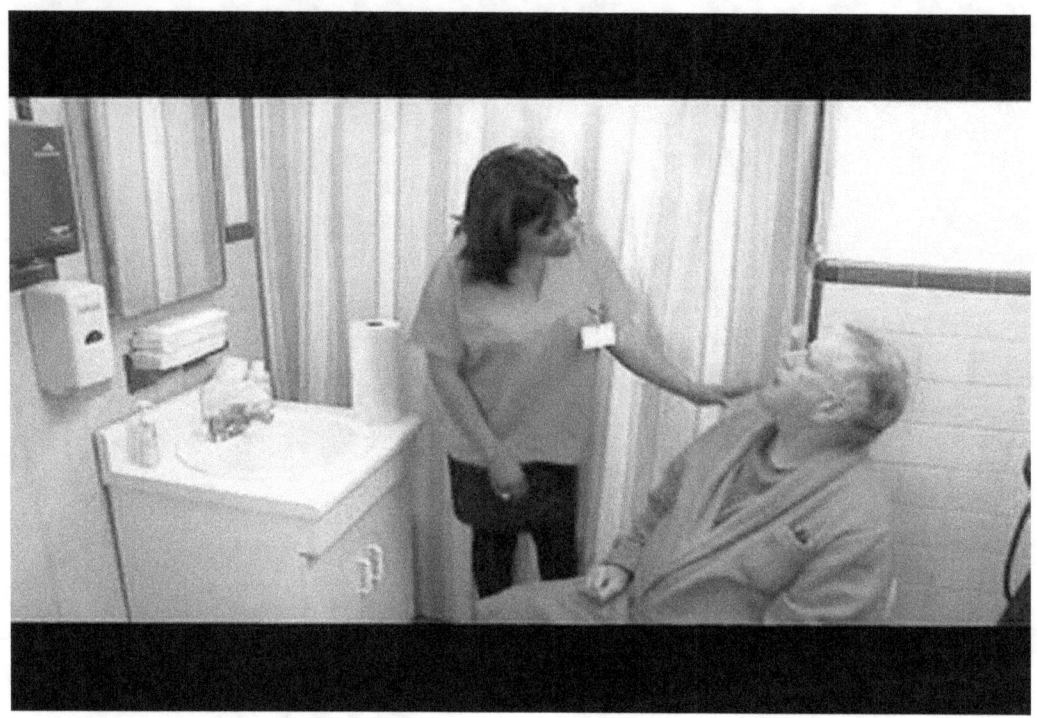

Case 5. Cultural Differences Around Physical Closeness

In this clip, touch—a form of non-verbal communication—is used by a Latino resident toward a care giver. The communication between the resident and care provider can convey a misunderstanding.

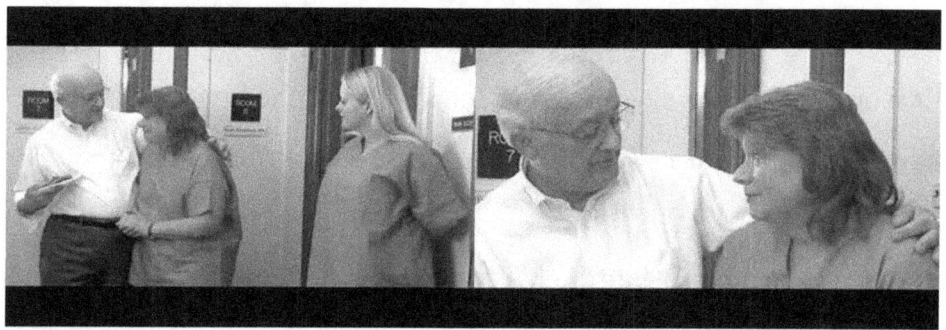

Activity: Active Listening

> *In groups of three, read each statement and the possible responses. Ask which responses show excellent or poor listening skills. Ask the group for other possible answers.*
>
> **1. Statement: "My daughter wants me to remain in the nursing home."**
>
> *Possible Responses*
>
> A. Well, maybe you should consider it. Things are hard for her at home.
>
> B. How could she say that! After all you've done for her!
>
> C. How do you feel about that?
>
> **The Answer is C, because it's a probing question to obtain more information and to understand feelings.**

Activity: Active Listening Continued

2. Statement: "I can't eat this food; it's boring!"

Possible Responses

A. Try this *Ensure*. I like it.

B. What's keeping you from eating?

C. Don't worry. You'll be sleeping soon, so you won't be hungry.

The Answer is B, because it's probing to gain more information

3. Statement: "I just can't handle Miss *A* anymore."

Possible Responses

A. Is there something that she did that is bothering you today?

B. I know what you mean. I have this resident who is driving me crazy, too!

C. The residents' families just don't care, and the hours here are long. You should just quit.

The Answer is A, because it's probing to learn what the real issues are.

4. Statement: "The music here puts me to sleep."

Possible Responses

A. I don't know why you ever listen to music.

B. You're upset about the music.

C. Everyone has to adjust to the situation here.

The Answer is B, because it's reflecting the speaker's feelings.

Activity: Active Listening Continued

5. Statement: "I can't understand anyone here. No one speaks English."

Possible Responses

A. I'm not surprised. The personnel director is non- English speaking.

B. Tell me about it! It angers me too.

C. What's going on that makes you feel that way?

The Answer is C, because it's probing for more understanding about the concern.

6. Statement: "Sometimes I feel like the residents want to do things their way."

Possible Responses

A. Oh, yes. They think they know more than anyone else around here.

B. It sounds as if you are feeling that the residents' concerns are important.

C. Well, it's not going to do you any good to feel sorry for yourself.

The Answer is B, because it's reflecting the speaker's feelings.

Refer to the voice-overs from the video for ideas to help facilitate discussion relating to this activity.

MODULE 4:
INTELLECTUAL COMPETENCE, INTERVENTION COMPETENCE, COGNITIVE STRATEGIES

Description

Communication skills are used in all patient activities, from taking health histories to assessing complaints. However, it is important to recognize that communication involves verbal as well as non-verbal elements. When there are differences in languages between the care provider and the resident, an interpreter may be needed to avoid misunderstandings, especially around important care issues. Non-verbal communication includes tone of voice, eye contact, physical distance and touch, emotional expression, and body movements. Another important component of cultural competency concerns learning how persons from various cultures understand their illness, and then finding ways to deal with different conceptual views of health. Finally, another aspect of cultural competency involves identifying and making interventions to enhance an organization's level of cultural competency.

Learning Objectives

After completion of this module you will be able to:

1. Provide assessments that show respect for individuals and cultures.
2. Describe appropriate steps in developing culturally appropriate verbal and non-verbal communication skills.
3. Describe suitable interventions on the individual and the organizational level for residents and their families in the nursing home.
4. To explore possible long-term and short-term goals for the nursing home.

Enhancing Verbal Communication—The Use of Interpreters

If care providers do not speak the same language as the resident, then professional interpreters should be used, especially around important care issues. In general, care providers should refrain from using family members, especially children, as interpreters. Although they may only want to help, such interpreters may not have the appropriate language skills in one or both languages. Moreover, family interpreters may be uncomfortable discussing some health issues that is perceived as sensitive. Care providers can be advocates for effective on-site interpreter services and for access to telephone based translation services. Professional interpreters are trained to create meaning when they translate rather than word-for-word translations. As expert interpreters, they make certain that complex concepts get across in both directions of the health related interaction. It is especially important to use a variety of strategies when doing care assessments.

Non-Verbal Communication

❖ *Pace of conversation*: Some cultures are comfortable with long periods of silence, while others consider it appropriate to speak before the other person has finished talking. Pay attention to cues from the resident and family members.

❖ *Physical distance*: Provide residents with a choice about what physical closeness they wish by asking them to sit wherever they like. Individuals from some cultures (e.g., American, European) tend to prefer to be about an arm's length

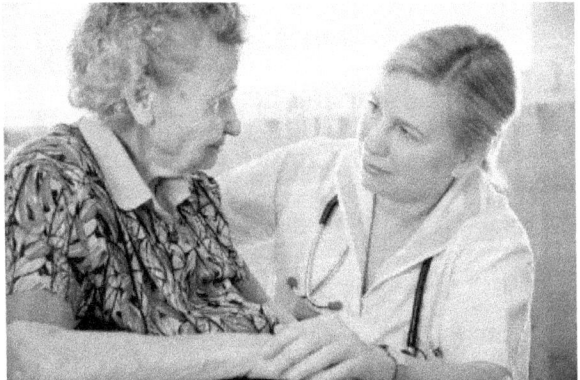

away from another person, whereas persons from some cultures prefer to be closer, e.g., some Hispanic cultures.

❖ *Eye contact:* While most European American ethnic groups typically encourage members to look people in the eye when speaking to them, some Blacks and Latinos may consider this disrespectful or impolite. Care providers are encouraged to observe the resident when talking and listening to get signs regarding proper eye contact.

❖ *Emotional expressiveness:* Some cultures value being unemotional, while others encourage open expressions of feelings such as sorrow, pain, or joy. Some Europeans tend to be more emotionally restrained, while some Caribbean and Latin American people tend to be more expressive.

❖ *Body movements:* Body gestures can be easily misunderstood depending on what is considered culturally appropriate in different parts of the world. Individuals from some cultures may consider some types of finger pointing or other typical American hand gestures or body postures as disrespectful or obscene, while others may consider strong hand shaking as a sign of aggression. In Western culture, an energetic hand shake is a sign of good will. When in doubt, ask the resident.

❖ *Touch:* While physical touch is an important form of non-verbal communication, the etiquette of touch is highly variable from culture to culture. Care providers should be aware as to whether touch is appropriate for cultures with whom they work, and they should feel comfortable expressing their own comfort or discomfort based on their own experiences.

The Use of Explanatory Models

The use of Explanatory Models for culturally competent care assessment is vital. An Explanatory Model is a belief system that people from a given ethnic or cultural group have about what has caused their illness and what the impact of the illness will be for them. Below, we describe various methods that have been found to be effective in improving communication and showing respect for the resident's point of view. These tools allow for the coordination of conventional strategies balanced with the cultural traditions of the resident. It is believed that if the model is used successfully, the residents' trust level increases. The goal of these assessments is to draw out residents' viewpoints on health conditions, and allows the residents to offer ideas for treatment. The following questions can be posed to residents as an example of an Explanatory Model.

A. What do you call your illness? What name does it have?
B. What do you think caused it?
C. When do you think it started?
D. What does your sickness do to you?
E. How severe is it? Will it have a long or short course?
F. What are the chief problems caused by your illness?
G. What do you fear most about your sickness?
H. What treatment should you receive? What are the most important results you hope to receive?

Strategies for Communicating Accurate Information

- Be sure you have the resident's attention before speaking.
- Use resident's name, pronounced correctly.
- State your name before making a statement or asking a question.
- Give concise information, while being sensitive to cultural issues.
- Speak slowly and clearly.
- Ask if you are being clear and be sure you are understood fully.

A care provider's thoughtful questions bring out more information from the resident and give him/her the opportunity to expand on a thought. In this way, the care provider can search deeper to learn more about health problems, and fears or joys with respect to the resident's culture. On the following page, you will see the acronym, *"LEARN."* It is a helpful technique to gather accurate and detailed information from residents.

Guidelines for Care Providers

"LEARN"

L *Listen* with sympathy and understanding to the patient's perceptions of the problem. Consider all residents as individuals first, and then as a member of a specific ethnic group.

E *Explain* your perceptions of the problem. Treat all facts as hypotheses to be tested. Turn statements of fact into questions.

A *Acknowledge* and discuss the differences and similarities. Remember all minority persons in this society are a blend of more than one culture. Identify strengths in the resident's cultural history.

R *Recommend* treatment and know your own attitude about the variety of cultures. Know whether you tend to stress the main health view or the traditional views.

N *Negotiate* agreement. Help residents understand the different views on health and reduce the negative feelings they may have of views different from their own. Keep in mind that there are no substitutes for good clinical understanding, caring, and a keen sense of humor.

Changing the Organization to Enhance Cultural Competency

Each nursing home as an organization must realize that working with residents from diverse ethnic backgrounds and their families is an active process. A balance between the care provider's cultural differences and residents and their families' expectations, continues to be a challenge. Strategies to promote cultural competency and sensitivity include:

1. Efforts toward maintaining ongoing staff development.

2. Sustaining open communication with residents and their families. Frequent involvement of family members on an individual level, as well as on a systemic level, improves the nursing home's services. A number of positive resulting factors can be expected such as a reduction in conflicts, a decrease in confusion, and a decline in the number of complaints. In the end, the nursing home will achieve greater resident satisfaction.

3. Hosting separate focus groups with all of the members of the nursing home community.

4. Using the family as a resource to improve cultural competent services in the nursing home. Less time should be spent handling disputes between staff and families when there is a commitment from both sides to support the resident. The organization should offer a variety of resources to families on a regular basis; an authorizing signature on care plans is not enough.

5. Engaging and supporting all family members at the nursing home. Open houses offered quarterly have been effective in increasing

residents' families support and in understanding the goals of the nursing home. Also, another important strategy is to encourage family members to participate in Alzheimer's Association support groups in the community.

The implementation of culturally competent care services is a process that cannot be rushed. Staff may show resistance and extra funding may be needed to implement new interventions. Time, training, and money are justified when a nursing home works diligently across cultures to honor and respect cultural beliefs, languages and traditions of its clientele. The outcome will allow for negotiation and compromise about cultural issues such as food, media options, music, religious and health practices. Nursing homes that are offering multilingual newspapers, more options with diet and entertainment, flexible visiting hours and accommodations for family members, have found that residents feel more satisfied.

Cultural Competency is on a continuum. A nursing home must determine where it is on the continuum. Through the process, staff should continue to examine their own beliefs and biases. When staff value differences, appreciate similarities, and become informed about cultural health beliefs and traditions, the nursing home as an organization, may still not be perfect. However, as an organization, it will be moving in the direction where the staff responds in a respectful manner to the residents. Improved relationships between care providers and residents will bring about greater satisfaction and a feeling of community within the facility.

Sources

Berlin, E.A. and Fowkes, W.C. (1983). A Training Framework For Cross-cultural Health Care. *Western Journal of Medicine* 139:934-938.

Green, James W. (1982). *Cultural Awareness in the Human Services*. Engelwood Cliffs, NJ: Prentice Hall.

Kleinman, A (1980) *Patients and Healers in the Context of Culture: An Exploration of the Borderland between Anthropology, Medicine, and Psychiatry*. Berkeley: University of California Press.

Video Case Studies

Case 6. An Organizational Goal

The clip shows a care provider expressing to an administrator an idea for change in the nursing home. The administrator is resisting. In the video voice-overs, possible alternatives are offered.

Activity: Organizational Change

> **List and discuss organizational/total changes you'd like to see put into action in your nursing home. Put these changes in two lists: short-term changes and long-term changes.**
>
> *Refer to the voice-overs from the video for ideas to help facilitate discussion relating to this activity.*

APPENDICES

Test Your Knowledge on Cultural Competence

Below is a list of questions regarding your knowledge of cultural competency. Please read each question below and decide the best answer.

1. **A care provider that is highly skilled in cultural competence:**
   a. Considers only the resident's health view when providing care.
   b. Knows about the resident's cultural traditions, values and customs
   c. Treats all cultural facts as absolute truths without testing them with residents.
   d. Considers all residents as individuals first, and then as a member of an ethnic group

The answer is D. Care providers who are highly skilled in cultural competence consider a resident as a unique individual first. Understanding the resident's cultural values will encourage respect, negation, and compromise about cultural issues.

2. **A care provider may not be attentive to residents' needs because:**
   a. Biases may stand in the way.
   b. Positive relationships at work exist.
   c. A resident's recognition is wanted.

d. The resident's family members are friendly.

The answer is A. Care providers must utilize a process of self reflection to analyze their own beliefs and biases that may hamper their involvement with residents. Being sensitive and respectful will help a care provider go beyond preconceived ideas of the resident.

3. **The following action is an example of a non-verbal communication:**
    a. Displays on a bulletin board.
    b. A pleasing smile.
    c. Giving a friendly "Hello."
    d. A written note to compliment a resident

The answer is B. The other choices incorporate words and are methods of verbal communication.

4. **The African American health belief system places significant value on:**
    a. Healing power of the prayer.
    b. Advanced medications.
    c. Modern medical surgical advancements.
    d. Relationships with saints and priestesses.

The answer is A. For many African Americans, there is a belief that a positive relationship to their faith and a belief that the strength of spirituality will maintain health and combat illnesses.

5. **A cultural competency model to identify the resident's view on illness will incorporate the following question:**
    a. What is the doctor's view?
    b. What are the resident's views about the illness?

c. How can medical science help?

   d. How can the resident satisfy the care provider's requests?

The answer is B. Residents from a given culture may have specific views on illness. It is important to ask for those views when coordinating and balancing conventional strategies with the traditions of the resident.

6. **The following behavior shows high intra-personal competence:**
   a. Sense of humor
   b. Stern behavior
   c. Impatience
   d. Distance from the resident

The answer is A. The art of providing good care to residents involves maintaining a sense of humor. It helps to " lighten up" the mood regarding the residents' conditions.

7. **African Americans and Latinos have a similar view with respect to:**
   a. Family
   b. Religion
   c. Health beliefs
   d. Dress

The answer is A. Both African Americans and Latinos place heavy emphasis on family and extended family members.

8. **The following statement reflects the cultural competency point of view:**

a. Although my professional or moral viewpoints may differ, I accept family and residents as the ultimate decision makers for services and support.
b. I accept that individuals from culturally diverse backgrounds hold the same views on culture.
c. I understand that family is defined as a mother, father and children.
d. I accept that religion and spirituality hold no importance in health issues.

The answer is A. Frequent involvement of family members on an individual level as a well as an on an organization level, improves communication, and reduces complaints from residents and family members.

9. An example of a care provider using an effective method to communicate to residents is:
a. Scolding residents.
b. Using non-verbal techniques, such as appropriate touch.
c. Complaining to supervisors.
d. Avoiding difficult residents.

The answer is B. It's important to use a variety of strategies when doing culturally care assessment. The strategy falls into the grouping of non-verbal communication. Other examples include eye contact, emotional expressiveness, pace of conversation, and physical distance.

10. It is important for care providers to know a variety of health beliefs in order to:

a. Make a decision about what health beliefs are more important.

b. Offer recommendations about health options.

c. Coordinate the understanding of various cultural perspectives.

d. Judge the value of each health belief system.

The answer is C. The care team looks to coordinate the perspectives of the residents, family members and conventional views. The goal is to determine whether the residents' health beliefs that are based on their culture, will do harm or good.

11. One of the main effects of *ethnocentrism* in the nursing home is:
   a. Improved understanding of differences in culture.
   b. The acknowledgment of similarities in ethnic groups.
   c. Creating open dialogues to foster respect.
   d. The continuation of stereotyping residents.

The answer is D. Ethnocentrism is seeing one's own culture as central and as the lens through which to view other cultures. A narrow, ethnocentric viewpoint leads to miscommunication, stereotypes, and possibly disrespect, which could lead to breakdown in culturally competent care.

12. The following statement regarding culture and traditions is generally true:
   a. Caucasians hold a health belief that relies on modern science.
   b. The African American family is structured around the nuclear unit.
   c. Latinos follow formal religions and believe less in native healers.
   d. Americans have one health belief system.

The answer is A. Choice A is the only true statement. The other statements are false.

Frequently Asked Questions about Cultural Competency

1. **Q. What does cultural competency care mean to care providers?**

   A. A care provider demonstrates high cultural competency skills when they show appreciation for similarities and differences in various cultures.

2. **Q. How does a care provider work to dispel stereotypes?**

   A. Care providers who engage in self-reflection, learn about various cultures, and participate in gatherings of groups with many cultures, are more likely to avoid using stereotypes.

3. **Q. What is a simple action that can be used with a resident to show care and appreciation?**

   A. A smile is usually returned with a smile.

4. **Q. What is one of the ways to value a resident's view of his/her illness?**

   A. It is important to ask the resident his/her view about the illness?

5. **Q. What are some ways to gather information about a resident?**

   A. Whenever you can, speak slowly to the resident, probe for more information, and never make assumptions about cultural values. If care providers do not speak the same language as the resident, then professional interpreters should be used. Care providers should be cautious about using family members as interpreters because it may affect the willingness of residents to speak their mind.

6. **Q. How can a care provider demonstrate active listening skills?**

A. The care provider should restate in their own words what the resident says.

7. Q. What are some qualities that show cultural sensitivity?

A. Some of the qualities that show cultural sensitivity are caring, openness, respect, acceptance, patience, and empathy.

8. Q. What are some ways the nursing home as an organization can improve cultural competency care?

A. Some suggestions that we have gathered in our work include the following: discussing culture as part of treatment for resident; offer options for food, music, and media; develop close relationships with family members; decorate bulletin boards that reflect the various cultures; and seek to have a staff that is representative of the various ethnic groups.

9. Q. How does a care provider become more self-aware of how his/her actions affect others?

A. Development of self-awareness guides the care provider to assess feelings of prejudice and discrimination. Some strategies ato increase self–awareness include: practice assertiveness and sensitivity; maintain a healthy sense of humor; keep a diary or notes on work experiences; stay in touch with your own feelings and get away from a difficult situation to remain calm.

10. Q. How can a care provider demonstrate high cultural competency skills?

A. Cultural competency skills are practiced continually. For example, when differences arise in staff values, also recognize and appreciate the similarities; become better informed about cultural health beliefs and traditions. The ultimate goal is to bring about a sense of community in the facility.

11. **Q. What are some effective methods to communicate to residents?**

    A. The care provider should use multiple methods including non-verbal techniques, such as appropriate touch; restate in your own words what has been said; and consult with family members and supervisors.

12. **Q. Why should a care provider know a variety of health beliefs?**

    A. The care provider is a professional and having knowledge about health beliefs allows for a better understanding of patients and their families, and for improved coordination of care between the health service team and the residents and family members.

Instructor's Post-Evaluation

Instructor's Name: _____

Participant's Name (optional): _____

Date: _____

Please check one box for each of the five questions:

	Item	Excellent	Good	Fair	Poor
		4	3	2	1
1	The instructor was well prepared.				
2	The instructor used your experiences.				
3	The instructor used discussions.				
4	The program met your needs at work.				
5	Overall, how would you rate this program?				

Additional comments: Please complete the statements:

1. **The session I found most helpful was (circle one):**

   a. Informational/Introduction

   b. Intra-personal Competence Self-Awareness

   c. Interpersonal/Communication

   d. Intellectual Competence /Intervention Competence

2. **The session I found least useful was (circle one):**

   a. Informational/Introduction

   b. Intra-personal Competence Self-Awareness

   c. Interpersonal/Communication

   d. Intellectual Competence /Intervention Competence

3. **I would recommend the program because:**

4. **Something I can use immediately is:**

5. **Things in the program that need to be improved:**

6. **I would suggest the following as ways to improve the program:**

RESOURCES

With the exception of Victoria Parker's program, the material presented in this manual has been adapted from work developed by programs in non-nursing home settings. The principal resources used to develop this program include:

Yeo, G., Hagan, J., Levkoff, S., Mackenzie, E., Mendez, J., Tumosa, N., Wallace, S. (1999). *Core Curriculum in Ethnogeriatrics*, Developed by the Members of the Collaborative on Ethnogeriatric Education. Bureau of Health Professions, Health Resources and Services Administration, U. S. Department of Health and Human Services.

Henderson, J.N, Henderson, L. C. (2003). *Cultural Competency for Practitioners Responding to Cognitive Impairment in American Indians*. Department of Health Promotion Sciences College of Public Health. University of Oklahoma Health Sciences Center Oklahoma City, Oklahoma.

Parker, V., Friedman, W., Hardt, E., Engle, R. Tabor, L. Lach, C. (2005). *Creating Solutions: Handling Cultural Complex Situations in a LTC Setting*, the Organizational Cultural Competence Research Team at Boston University School of Public Health. Video produced by *Better Jobs Better Care,* a national program supported by the Robert Wood Johnson Foundation, the Atlantic Philanthropies, Future of Aging Services and American Association of Homes and Services for the Aging (AAHSA).

Cervantes, E, Heid-Grubman, J., Schuerman, C.K. (1995). The *Paraprofessional In Home Health and Long-Term Care, Training Modules for Working with Older Adults*. The Center for Applied Gerontology, Chicago, Health Professional Press.

Adams, A., Walcott-Brown, L. (2001). *Alzheimer's Training Curriculum for Direct Health Care Providers,* Brooklyn Alzheimer's Disease Assistance Center (BADAC), SUNY Downstate Medical Center, Brooklyn New York.

www.ingramcontent.com/pod-product-compliance
Lightning Source LLC
Chambersburg PA
CBHW080818170526
45158CB00009B/2464